EXECUTIVE SUMMARY

Beyond Pesticides

BIOLOGICAL APPROACHES TO PEST MANAGEMENT IN CALIFORNIA

DIVISION OF AGRICULTURE AND NATURAL RESOURCES   •   UNIVERSITY OF CALIFORNIA   •   1992

For additional copies of *Beyond Pesticides: Biological Approaches to Pest Management*, please send $7 for each executive summary (ANR Publications Number 21512) and $14 for each book (ANR Publications Number 3354) to:

ANR Publications
University of California
6701 San Pablo Ave.
Oakland, CA 94608-1239

Please specify the publication numbers in your order. Make checks or money orders payable to UC Regents. Price includes postage and handling.

You may also order by phone or FAX, using the following numbers:

ANR Publications, telephone: (510) 642-2431
ANR Publications, FAX: (510) 643-5470

ISBN 1-879906-10-4

Library of Congress 92-061146

All photographs in this executive summary are by Jack Kelly Clark, UC Davis, unless they are otherwise credited.

Cover photographs:
Large photo: Plastic mulch suppresses weeds and reduces water use in these young strawberry fields in Santa Cruz.

Small photo, left: The Hyposoter wasp, *Hyposoter exiguae,* is laying its egg in a beet armyworm. Many species of tiny wasps parasitize and kill caterpillars.

Small photo, center: A UC farm advisor shows growers how to use a handlens to check for the presence of pests and natural enemies.

Small photo, right: The fungus *Arthrobotrys dactyloides* captures nematodes with constricting rings. One constricting ring is unsprung, the other has constricted around a nematode that moved into the trap. (Photo: Bruce A. Jaffee)

Title page photograph:
Adult syrphid flies feed on pollen and nectar; however their larvae are important aphid predators.

EXECUTIVE SUMMARY
CONTENTS

FOREWORD

SCIENTISTS ARE OFTEN SEEN AS pathfinders with eventual "destinations"—whether that means developing an AIDS vaccine, saving the ozone layer, or launching a human colony into space. In some research, however, there is no dramatic end in sight. For instance, the battle against crop pests is eternal, with only momentary periods of relief when science temporarily gains the advantage. New pests are continually introduced despite efforts to regulate plant and animal importation. Existing pests develop resistance or evolve into new, more destructive forms. Few people realize that our "scientific advantage" in agriculture has been dwindling rapidly—and pest problems loom as large today as they did 100 years ago.

Once armed with an arsenal of synthetic pesticides, growers now find their weaponry is becoming obsolete due to pest resistance, or is simply disappearing from market shelves due to health and environmental concerns. As existing legislation is implemented over the next five years, California farmers are expected to lose more than 150, or 50%, of the active ingredients in current agricultural use, the result of manufacturer decisions not to pursue re-registration. New pesticide development can cost as much as $70 million and require years of research. In today's market, manufacturers seldom assume this expense

unless a new pesticide can be registered for at least one "major crop" (usually corn, wheat, soybeans or cotton). Many "minor use" pesticides—once indispensable tools in producing California's cornucopia of specialty crops—have now

Pheromone dispensers are placed in a pear orchard—a biological approach to codling moth control. So much sex pheromone is emitted that male moths cannot find mates.

been lost and not replaced, leaving farmers to seek alternatives which may not exist or for which effectiveness is uncertain.

This report is not based on the assumption that all pesticides will or should be eliminated. Rather, it recognizes that most farmers, like society at large, want to reduce pesticide use, both because of its increased cost (as much as 20 percent of production for some specialty crops) and its decreased effectiveness. For the last decade, California growers have begun to shift away from synthetic pesticide use to practices that reduce potential harm to humans and the environment. This has been possible, in part, because UC scientists have made impressive gains in integrated pest management, biological control and genetic engineering. Unfortunately, the demand for pesticide alternatives has far outstripped their actual rate of development.

Although developing effective alternatives depends on research, investment in the research structure that once made U.S. agriculture the envy of the world

has been declining for three decades. The federal government's spending on agricultural research decreased from 13% of all non-defense R&D in 1955 to only 4% in 1990. At the state level during the last decade alone, the combined effects of inflation and escalating costs of instrumentation to conduct increasingly complex research have resulted in a severe erosion, perhaps 50% to 60% in real terms, in the non-salary components of the UC Agricultural Experiment Station (AES) budget. In the last two years alone, budget cuts have slashed state AES research and public service funding 10%.

As a result, applied research efforts have plummetted relative to basic investigations, and basic research has fallen short of filling many important gaps in our knowledge. For example, in 1990 UC scientists analyzed more than 600 crop-pest situations and found alternatives to synthetic pesticides now exist for 75% of them. However, fundamental questions about how these methods work, their practical potential, and their

Since its founding in 1959, UC's Lindcove Field Station in Tulare County has provided California growers with citrus cultivars which are higher yielding, longer living and disease-resistant.

economic feasiblity are largely unanswered—awaiting further research.

It is particularly ironic that the deterioration in support for research has occurred at a time when society's concern about environmental and conservation issues has increased. The future of California agriculture, wildlands, and forests may now depend on the rapidity and effectiveness with which actions are taken to hasten the development and application of effective pesticide alternatives.

Beyond Pesticides: Biological Approaches to Pest Management is the first in-depth analysis of biological pest management under California's diverse climactic and crop conditions. It assesses research accomplishments and failures, identifies crop-pest situations most amenable to biological approaches, and lists research imperatives in each pest management discipline. It is the product of a two-year effort by a group of internationally recognized University of California scientists. The book has been designed to be a working guide for scientists nationwide seeking to maximize their efforts in this vital area of research.

The book consists of an executive summary and seven chapters. Each of the first six chapters explores the biological management of one type of agricultural pest; the seventh discusses resource and policy issues which must be addressed if we are to encourage adoption of biological approaches.

For the reader's convenience, the executive summary has also been printed separately. It includes an introduction, general research imperatives, and major findings and recommendations. The separate version of the summary also contains abstracts and specific

Farmers' markets offering organic produce have sprung up all over the state in recent years.

research imperatives from each chapter of the book.

If scientists are to help farmers find adequate and safe pesticide alternatives, they must integrate knowledge across disciplines in the context of the state's problems. UC's Division of Agriculture and Natural Resources is actively committed to meeting California's urgent need for biological approaches to pest management. We offer this document in the hope that its contents will shape the course of future research by UC faculty and their partners in the public and private sectors.

Kenneth R. Farrell
Vice President
Division of Agriculture
and Natural Resources
University of California

ACKNOWLEDGMENTS

THE STUDY GROUP BELIEVES THIS document represents a thorough review of the opportunities and challenges lying ahead. We hope it will serve as the foundation for the next generation of agricultural research in California. We envision an era of research dedicated to the goal of moving beyond pesticides, research that makes biological approaches to pest control increasingly productive and profitable. To the extent that we succeed, future generations will reap rich benefits, through an agriculture that is not only abundant, but also beneficial to the environment and safe to farm workers and consumers.

A debt of gratitude is owed to those who contributed to the completion of this report. During the past 18 months, we have called upon dozens of University of California faculty and U.S. Department of Agriculture ARS scientists to help produce an historic document. In addition to the Study Group members, we have listed the principal and contributing authors for each chapter. Many others contributed additional material and are, we hope, properly acknowledged in the book length version of this report.

The Study Group on Biological Approaches to Pest Management

J. Patrick Madden
PROJECT DIRECTOR AND GENERAL EDITOR, OFFICE OF THE VICE PRESIDENT OF UC, DANR

Milton N. Schroth
PLANT PATHOLOGY, UC BERKELEY, CHAIR OF STUDY GROUP

Thomas C. Baker
DEPARTMENT OF ENTOMOLOGY, UC RIVERSIDE

Harold O. Carter
AGRICULTURAL ISSUES CENTER, UC DAVIS

Donald L. Dahlsten
DEPARTMENT OF ENTOMOLOGY, UC BERKELEY

Howard Ferris
DEPARTMENT OF NEMATOLOGY, UC DAVIS

Mary Louise Flint
UC STATEWIDE IPM PROJECT, UC DAVIS

James M. Lyons
DIRECTOR, CENTER FOR PEST MANAGEMENT RESEARCH AND EXTENSION, OFFICE OF THE VICE PRESIDENT OF UC, DANR

Richard H. McCapes
DEPARTMENT OF EPIDEMIOLOGY AND PREVENTIVE MEDICINE, UC DAVIS

Michael L. Morrison
DEPARTMENT OF FORESTRY AND RESOURCE MANAGEMENT, UC BERKELEY

Michael K. Rust
DEPARTMENT OF ENTOMOLOGY, UC RIVERSIDE

Charles E. Turner
U.S. DEPARTMENT OF AGRICULTURE, AGRICULTURAL RESEARCH SERVICE, ALBANY

Patrick V. Vail
U.S. DEPARTMENT OF AGRICULTURE, AGRICULTURAL RESEARCH SERVICE, FRESNO

Frank G. Zalom
UC STATEWIDE IPM PROJECT DIRECTOR, UC DAVIS

Executive Summary Authors

Foreword:

Kenneth R. Farrell
VICE PRESIDENT FOR AGRICULTURE AND NATURAL RESOURCES, THE UNIVERSITY OF CALIFORNIA

Executive Summary:

Mary Louise Flint, James M. Lyons, J. Patrick Madden, Milton N. Schroth, Albert R. Weinhold, Janet L. White, and Frank G. Zalom, with Mary Jean Haley

BEYOND PESTICIDES: HIGHLIGHTS

❈ California agriculture generates about $57 billion in direct and indirect revenues each year and provides one in six jobs in the state. It is an important source of food and fiber for the state, the nation and the world.

❈ Numerous pesticides that have made the style and scale of California agriculture possible are being lost— some rendered obsolete by pest resistance, others withdrawn for posing a serious threat to human health or the environment.

❈ Meanwhile, new pests—recently arrived exotic species or newly evolved pesticide-resistant strains—pose an unrelenting challenge to agriculture.

❈ Public concerns about the effects of widespread pesticide use on worker safety, the health of the general public, and the environment are generating a growing pressure to find safer, more eco-logically sound alternatives to synthetic chemical pesticides.

❈ One of the most promising means to alleviate California agriculture's dependence on synthetic chemical pesti-cides is to expand biological approaches to pest control.

❈ If biological approaches are to reach their full potential, scientific progress is needed in five areas:

1. The development and mainte-nance of host plant resistance

2. The introduction and augmen-tation of natural enemies as pest antagonists

3. The improved understanding of the ecology of pest populations, biological control agents, and plants in agricultural systems

4. The development of behavior-modifying chemicals (semio-chemicals) as control agents

5. The development of auxiliary research such as pest population monitoring and determination of economic thresholds

❈ University of California expertise is a state resource vital to the further development and wide adoption of such research in California. UC scientists have pioneered biological approaches to pest management over the last century.

❀ Yet as pest problems and environmental and public health concerns have grown, UC budgets for addressing them have dwindled. Current funding for research into biological pest management must increase to meet the expanding needs of the state's agricultural industry.

❀ Research on the social and institutional barriers to the adoption of biological pest control strategies is also essential. Special attention should be devoted to the economic risks growers face during the transition period as they experiment with new practices.

❀ The development of viable pest management alternatives is a reachable goal. It requires immediate and decisive action by California state legislators to support the University of California's accelerated development of biological pest management approaches, and their rapid adoption by growers statewide.

Herbicides like the one being applied to this field account for 60% of the pesticides sold in California.

OVERVIEW

CALIFORNIA AGRICULTURE GENERates about $19 billion at the farm gate each year and more than $38 billion in related economic activity. It provides one in six jobs in the state. California farmers produce over 250 different crops and livestock commodities and are the nation's leading producers of 58 of them. In addition, one in four California farm acres is devoted to growing agricultural exports —making California agriculture important not only to the state and nation, but to the world.

It is the purpose of this report to point out that an urgent threat to this significant industry has been identified and to suggest ways of dealing with it. It is the conclusion of this report that the future of California agriculture, along with the health of the state's inhabitants and its environment, requires an immediate major investment in developing biological approaches to pest manage-

Note: A summary is necessarily abbreviated and incomplete. The Study Group on Biological Approaches to Pest Management offers this Executive Summary as a convenience to its readers and an introduction to its work. However, the Study Group recommends reading the entire Beyond Pesticides report to obtain a comprehensive understanding of the subject matter. (See page 2 for ordering information.)

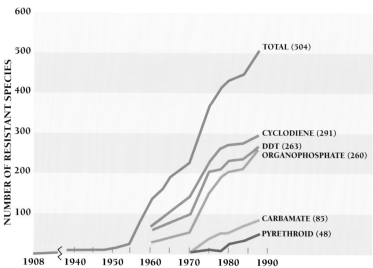

ment as a practical alternative to synthetic pesticides.

Growers statewide have lost a large number of the chemical weapons that have made the style and scale of California agriculture possible. They could well lose over 150 more in the next five years as existing legislation is implemented to reduce pesticide risks to human health or the environment. Others are becoming obsolete as pests develop genetic resistance to them (see figure 1). The loss of these compounds represents the steady erosion of a key tool that has helped make California agriculture the economic powerhouse it is today.

California decision makers need not be helpless in the face of emerging pest

Figure 1. The number of species resistant to major classes of pesticides has risen dramatically in recent years. (Source: *The Occurrence of Resistance to Pesticides in Arthropods*, George P. Georghiou and Angel Lagunes-Tejeda, U.N. Food and Agriculture Organization, Rome, 1991.)

Although pesticides have been California's dominant pest control agent since the 1940s, their use is increasingly restricted because of health and environmental regulations and the development of pesticide resistant pest strains.

control problems. While many forces that shape the state's agriculture are largely or entirely beyond state control—the prolonged drought, the emergence of new or chemical-resistant pests, and federal regulatory decisions on pesticide registration—the development of viable pest management alternatives is a reachable objective. It requires decisive action to make a long-term commitment of new public funds to accelerate the development and adoption of biological pest management approaches.

What are "biological approaches to pest management?" They encompass a variety of innovative tactics. An example is the use of microbial insecticides to cause fatal diseases in specific pests. Another is to modify plants genetically to increase their resistance to disease or insect attack. Yet another is the use of insect sex pheromones to confuse and frustrate mating in crop-destroying moths. Biological approaches also include the use of environmentally safe chemicals such as botanical insecticides,

The microbial insecticide *Bacillus thuringiensis* (Bt) that killed this caterpillar was the first and is now the most widely used microbial insecticide in the world. Its early development and rise to commercialization were largely the result of the pioneering efforts and vision of UC Berkeley insect pathologist E.A. Steinhaus in the 1940s.

BRIAN FEDERICI LABORATORY

Bt, a single-celled bacterium, produces a toxin (the clear oval shape) that causes fatal disease in specific insect pests.

or the practice of crop rotation to prevent pest build-up in the field. Some biological control approaches, such as classical "biological control," are almost as old as agriculture, while others are just emerging from the lab.

University of California scientists have provided international leadership in advancing such methods over the last century. They pioneered biological control through foreign exploration to find and import the natural predators and parasites of pests; they bred disease-resistant crop plants and elucidated the mechanisms of insect resistance. In the 1950s, UC scientists advanced the principles of integrated pest management (IPM), and in the 1970s and 1980s, they generated basic knowledge that led to the development of semiochemicals (such as insect sex pheromones) and genetic engineering techniques. While such success illustrates the feasibility of biological approaches to pest management, the overall magnitude of development

ments in this area is small in relation to the need and the opportunities. The University of California's scientific expertise and landmark investigations have placed it in a unique position to foster more extensive development and wide adoption of biological pest management techniques.

Largely because of enormous investments in research and institutional support that were made many years ago, crop losses from pest damage have been kept in control in recent decades. Today's supermarket shopper enjoys year-round abundance but gives little thought to the farmer. This complacence is unjustified. The recent devastation of Southern California crops by strain B of the sweetpotato whitefly is a good example of the rapidity with which new pest control problems can develop. Despite widespread insecticide spraying, this pest is expected to cause hundreds of millions in crop damage annually and throw thousands out of work. This pest

In 1923, with the establishment of the biological control facility at UC Riverside, UC became the first university in the country to have a special biological control unit. This photo shows the facility in about 1940.

is not only resistant to many pesticides, it infests and damages an unusually wide variety of crops.

All indications are that California agriculture will continue to be vulnerable to new, chemical-resistant, and destructive pests. Attempts to control them with synthetic pesticides may be far too costly in environmental and economic terms, and in some cases, completely ineffective. Biological approaches, on the other hand, rarely raise the issues of environmental damage or threats to human health posed by conventional chemical controls and, in some cases, once they are successfully in place, they continue to work without further effort by, or expense to, the grower.

While biological pest control methods have provided some dramatic successes in the past, this area of research has not received funding commensurate with needs. In fact, funding to develop alternative pest control methods has been declining even as pest problems have grown, and it is now inadequate to meet the current needs of the state's agricultural industry, not to mention those of the future.

What is a pest?

Agricultural ecosystems are artificial, and they need constant maintenance. Crop pests are always present, always evolving, and always pose a potential threat to specific crop or livestock enterprises. Although the most prevalent image of an agricultural pest is probably the voracious insect, damaging pests also take the form of weeds, nematodes, plant and animal disease agents, and even vertebrates such as rodents and coyotes. The dollar value of crop losses

The pesticide-resistant strain B of the sweet-potato whitefly, which is shown here, unexpectedly devastated cotton and vegetable crops in the Imperial Valley in 1991, demonstrating how rapidly new and serious pest control problems can develop.

caused by weeds, for example, is estimated to be about a billion dollars a year in California. Herbicides, the primary tool used to combat weeds, make up over 60 percent of the pesticides sold in California (see table 1).

Various plants, animals, and microbes become "pests" when they cause economic, health, or aesthetic problems for humans. This occurs only at particular times and places because of changes in population size or in behavior patterns that result from natural or human-induced alterations in the environment. Almost every economically important insect and mite pest in California has become resistant to at least one pesticide, as have many plant pathogens. Many pests have become resistant to several types of pesticides (see figure 1). In some areas, the available chemical materials can no longer control any of the major pests of some crops. Even if environmental and health concerns were not a factor, a greatly increased effort to expand the arsenal of biological controls would be needed.

Synthetic chemical pesticides: a brief history

Although it sometimes seems they have always been with us, synthetic chemical pesticides have a fairly brief history. Highly toxic materials such as those based on arsenic or lead had been used on high value crops since the nineteenth century, but the synthetic organic insecticides were not developed until World War II. Once introduced, however, their use increased rapidly.

For the past 40 years, most California growers have preferred to use pesticides to control crop pests. There have been a number of reasons for this preference. The effect of pesticides on pest populations has often been immediate and dra-

matic. Pesticides have been relatively easy to obtain and use. Growers have considered them to be inexpensive, partly because environmental and health costs were neither well understood nor borne entirely by the pesticide users. Through the use of pesticides, growers have been able to produce some crops profitably in otherwise unsuitable locations. Pesticides have made feasible the monoculture of genetically similar crops (which have genetically similar vulnerability to pests), simplifying the management of large operations. Pesticide use

Consumers could soon find that California's abundance can no longer be taken for granted.

has enabled farmers to extend growing seasons and thereby take advantage of valuable market niches. Finally, pesticides have been widely used after harvest to maintain product quality and extend shelf life.

Since the 1970s, the trends in pesticide use have been mixed, with usage increasing in some categories, notably herbicides, and decreasing in others such as insecticides (see figure 2 and table 1). In some crops, the adoption of

Figure 2. Agricultural Pesticide Use in the U.S.
Pesticide use has begun to level off and may be showing a downward trend. Usage patterns are affected by changes in pest biology (including the development of resistance to pesticides), and legislative and regulatory action following the discovery of negative effects on water quality and the environment. (Source: Environmental Protection Agency.)

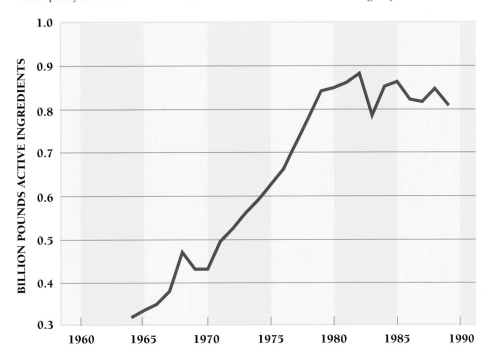

Table 1. U.S. Annual Volume of Pesticide Usage by Type, 1979-1989. (Source: EPA/OPP/BEAD estimates.)

| Millions of pounds of active ingredients used, by year | | | | | | | | | | |
Pesticide Type	1979	1980	1981	1982	1983	1984	1985	1986	1987	1988	1989
Herbicides	560	555	570	544	575	675	670	655	645	660	655
Insecticides	378	395	405	391	255	270	300	295	260	268	226
Fungicides	106	120	123	119	68	80	82	86	122	132	111
Other	106	105	107	106	55	55	60	60	60	70	78
TOTAL	1,150	1,175	1,205	1,160	953	1,080	1,112	1,096	1,087	1,130	1,070

integrated pest management (IPM) and biological controls has helped reduce pesticide use sharply.

Biological approaches: a longer, more obscure, history

As we have noted, biological strategies for pest management include methods that employ another organism as a control agent, any approach that disrupts natural behavioral patterns through the application of behavior-modifying chemicals (semiochemicals such as sex pheromones), the use of pest-resistant or pest-tolerant crops, and cultural practices that interfere with the normal life cycle of the pest organisms.

The key difference between biological and chemical approaches to pest management is selectivity. Most conventional pesticides affect a broad spectrum of living things. Because biological approaches target specific biological vulnerabilities of pest species, most nontarget species are left unharmed. As a result, the environmental and human health objections to conventional chemical pesticides do not apply to biological control approaches.

Unlike chemical pesticides, biological strategies, broadly defined, have a very long history, probably stretching back to the beginnings of agriculture. What is known as "classical" biological control, the deliberate introduction of natural enemies to control a pest population, was well developed by the late nineteenth century. The control of the cottony cushion scale in Southern California citrus groves in 1888-89 was the first major success of a classical biological control program in California. Since then, the University of California has been involved in developing more than 30 successful examples of this type of control.

One irony is that classic biological control is often undervalued because it is

One excellent example of long-term biological control is offered by the red and black vedalia beetle (left), which has protected California's highly productive citrus groves from damage by the cottony cushion scale since 1889.

Lindcove Field Station maintains the Citrus Clonal Protection Program foundation block, the source of disease-free budwood used by citrus growers throughout the state.

almost unnoticeable when it works best. In Southern California coastal citrus orchards, for instance, pests are often managed by biological control, with only occasional use of selective pesticides. Over the last century, citrus whitefly, black scale, yellow scale, red mite, snails, and mealybugs have been managed by parasitic wasps and predatory snails, insects, mites and spiders. In some cases a biological control agent must be reintroduced or reapplied from time to time, but in others the control agent has become established, and the two species live on in the field in balance. The presence of the formerly destructive pest, and its predator, are forgotten.

UC scientists have advanced biological approaches across a broad range of disciplines, from microbiology to weed science. Work done at UC includes further development of the microbial insecticide *Bacillus thuringiensis* (Bt) and recent research involving the use of pheromones and other semiochemicals in pest management. Sex pheromones have been used to create "better bug traps" capable of monitoring pest populations accurately, thereby reducing the need for prophylactic pesticide spraying.

Ecological and biological pest control principles are applicable throughout nature. However, the task of developing and applying biological control approaches is complicated by environmental factors, such as variations and extremes in rainfall and temperature, on both pest and control agent. Given the extremely large numbers of species that are or could become agricultural pests and the many possible ecosystem perturbations, it is easy to understand why

The results of basic research become useful to growers most quickly and effectively when UC Cooperative Extension personnel are involved in its evolution and delivery to the field.

COURTESY OF R. D. GOEDEN

progress in some areas has been slow. There have, however, been spectacular successes which provide the incentive and stimulus to continue research.

No overnight solutions

Scientific research is not a stop-and-start proposition. It builds from year to year as researchers begin to understand the complexity of the systems they study. In addition, there is an important distinction between basic (fundamental) and applied (problem-solving) research. A common rule of thumb is that there is typically a decade between a discovery at the laboratory bench and its development into something that can be applied at the practical, field level.

Because many interrelated biological and physical processes are involved in protecting the world's supply of food and fiber, developing and implementing biological methods of pest management must, of necessity, take a broad and interdisciplinary approach. There must be involvement of both university-based basic and applied researchers, including entomologists, plant pathologists, taxonomists, nematologists, crop scientists, economists, and animal health scientists,

to name a few. Information must flow in both directions along a continuum ranging from molecular biology to applied ecology and practical application on commercial farms. In addition, county Cooperative Extension personnel, who further develop the research, must be fully integrated into the information flow along with growers, foresters, and caretakers of reserve lands near agricultural areas. The viability of this network relies upon the continuous presence of qualified, dedicated personnel with adequate resources to identify and investigate problems and apply solutions in the field. The pest control problems facing California agriculture can't be solved overnight by massive hiring after serious crop losses have become common, or by focusing exclusively on a single approach or discipline.

The Center for Pest Management Research and Extension

If research is to be efficient and used to its full potential, basic and applied efforts must be coordinated. Programs must be carefully structured to ensure that researchers in the field and in the laboratory are communicating with each

A long-term biological control program has been very effective in controlling weedy prickly pear cactus on Santa Cruz Island. At left is a prickly pear stand in October, 1964. The same area is shown on the right as it was in September, 1980. The principal biological control agent used was a cochineal insect (*Dactylopius* sp.) that is native to the Southern California mainland.

18 ❀ BEYOND PESTICIDES

Scientists use molecular biology techniques to identify desirable genes such as those for plant disease resistance. Basic research plays an important role in developing most new pest management technologies. It can take a decade or longer for the fruits of fundamental research to reach the field.

An IPM advisor discusses leaf removal techniques with Sonoma County growers. Such methods have reduced pesticide use in wine grapes.

other. Technology transfer programs should be monitored to make sure that growers and other land managers are receiving the information they need to use new discoveries effectively. Finally, information about the uses and potential of new discoveries must be disseminated to policy makers and the general public.

In recognition of these needs, the Vice President of the University of California

Division of Agriculture and Natural Resources (DANR) established the Center for Pest Management Research and Extension (CPMRE) on April 1, 1991. CPMRE is helping to coordinate pest-related research, implement important new research findings, and integrate research findings with policy making and regulatory activities in the state. This center is designed to have a strong outreach program and to work closely with state policy makers.

As a division of the University of California, DANR currently brings together nearly 1,100 research scientists and educators on three UC campuses (Berkeley, Davis, and Riverside), nine field stations, and 64 Cooperative Extension county offices to develop and deliver practical solutions for local problems. Their efforts range from technical farm assistance and research in water conservation to innovations in veterinary medicine. The Division's three major components are the Agricultural Experiment Station, Cooperative Extension, and the Natural Reserve System.

GENERAL RESEARCH IMPERATIVES

IN APPRAISING THE STATUS OF BIOLOGical approaches to pest control in California, and in consulting with many faculty, the Study Group has developed the five general research imperatives presented here. These imperatives are concerned with those areas in which scientific progress is needed if biological approaches to pest management are to reach their full potential. Additional specific research imperatives and recommendations for major pest control problem areas drawn from each of the seven chapters of *Beyond Pesticides* are presented at the end of this summary.

1. RESEARCH IMPERATIVE

Increase research emphasis on the development and maintenance of host plant resistance.

Host resistance (the inbred ability of a plant to resist a pest or pests) is the control approach of choice whenever it is available because it is effective, economical, easy to use, and has no harmful effect on the environment. Since host resistance can apply to virtually every pest group, enhanced research support would be valuable in all problem areas. Biological research, including both laboratory and field studies, offers many novel and potentially valuable strategies to extend the usefulness of host resistance. The successful development of these strategies will require the fostering and

STEVE KOIKE

STEVE KOIKE

Planting pest resistant crop varieties is one of the easiest, most economical biological approaches to pest management. The 'Bossanova' spinach cultivar (left) has proven resistant to a new race of downy mildew. The widely-planted 'Polka' cultivar (right) is susceptible to this new pathogen.

support of balanced, cooperative efforts along a continuum ranging from molecular biology to applied ecology and practical application on commercial farms.

2. RESEARCH IMPERATIVE

Significantly expand research on introducing and augmenting the natural enemies and antagonists of pests.

Most examples of successful biological control involve the use of the natural enemies and antagonists of pests. Many opportunities remain to be exploited. There is a definite need for three types of modern facilities and trained personnel to staff them. The first need is for more and better quarantine facilities to increase researchers' ability to import, study, and rear exotic biological agents safely. The second need is for insectary facilities that will allow researchers to study and improve techniques for mass rearing and release of natural enemies. The third is for maintaining reference collections of pests and natural enemies. These collections are essential for correct identification and a key element in biological control and other biological approaches to pest management.

3. RESEARCH IMPERATIVE

Enhance our understanding of the ecological base of the populations of pests, biological control agents, and plants in agricultural systems and the interactions among them.

In moving beyond pesticides and recognizing the limitations of resistance,

biological approaches to pest management offer a major hope for managing a large number of production-limiting plant pathogens, weeds, nematodes, insects, mites, and vertebrates. A broad spectrum of pests could be controlled through the use of naturally occurring biological control agents, and through host plant resistance and various cultural manipulations such as crop rotation, mechanical cultivation, and cover crop management.

A better understanding of the ecological interactions of plants, pests, and natural enemies in the field is essential to using naturally occurring

A tiny *Aphidius* parasitic wasp lays its egg in a blue alfalfa aphid. Although the use of biological control agents of this sort has been very successful in some crops, the enormous potential of biological control still remains largely untapped.

biological control agents effectively, to enhancing the impact of cultural manipulations on pest populations, and to devising new cultural and biological controls.

Developing this knowledge base will require basic investigations of organismal, population, and community ecology, combined with field studies based in the varied production scenarios of California agriculture. There is also need for studies on genetic traits that both limit and enhance the ecological fitness of biological control agents. Again, cooperation among basic scientists, applied researchers, and innovative growers is essential.

4. RESEARCH IMPERATIVE

Continue and accelerate research into the development and use of semiochemicals.

Semiochemicals are behavior-modifying substances, such as pheromones or mating attractants. Both basic and applied research are required to exploit the opportunities semiochemicals offer. Their use is a rather specific approach to pest control that is largely restricted to arthropods (especially insects and mites) and nematodes. Nevertheless they are extremely important and, when applicable, have many advantages. Important research areas include the study of semiochemical-mediated insect and

UC researchers collect data on tomato yield and insect populations. Such information has enabled Sacramento Valley growers to rely on biological control of the tomato fruitworm at certain times of year.

nematode behavior, the identification of semiochemicals, and the development of improved field application methods. Despite the enormous potential for specific semiochemical approaches to disrupting pest behavior, very little is known about them at a molecular or physiological level. This basic knowledge will be essential to major scientific breakthroughs in the future.

5. RESEARCH IMPERATIVE

Support research in areas auxiliary to direct biological pest management approaches. These areas include systematics (concerning ecological relationships of organisms), monitoring pest populations, and determination of economic thresholds (levels at which pest populations reduce profits).

Developing effective biological approaches to pest management requires that pest and beneficial species be identified correctly and their phylogenic and ecological relationships be clearly understood. Taxonomy (classification and identification of organisms) and systematics (a subdiscipline of taxonomy dealing with the ecological relationships among organisms) are essential for identifying and understanding the biology of pests and their natural enemies. Systematics and taxonomy have a direct and crucial application to biological approaches yet are distinct specializations that must be supported as such.

Monitoring pest populations is another integral part of research and implementation in biological pest

UC scientists are performing basic research to identify communication chemicals (pheromones) produced by many insect pests, such as the stink bug shown here.

A pest control advisor checks a strawberry field for pest damage. Biological control approaches work best when growers or their pest control advisors closely monitor the effectiveness of the control methods they are using.

management approaches. Further development of monitoring techniques is essential and will require research on the distribution and abundance of pest populations and their natural enemies at high and, particularly, low densities. Applied ecological studies in actual production situations are of immediate and paramount importance. Research in promising fields such as molecular biology and immunochemistry may also prove to be essential.

The determination of economic thresholds encompasses the field of crop loss prediction. The specific methodologies involved are closer to quantitative epidemiology than to organismal biology. This field must be supported to give growers the information they need to make informed decisions about the economic and environmental implications of biological versus synthetic chemical pesticide approaches to pest management. Intensified research and development must be focused on improving and documenting the performance of new and better biological approaches to pest management. Only then can California move beyond pesticides into a new era of ecologically safe, productive, and profitable agriculture for present and future generations.

IMPLEMENTATION

PERFORMING THE RESEARCH THAT IS necessary to develop practical, effective, and widely used biological pest management approaches will naturally require a commitment of resources, but fortunately, the university already has at least one very successful program to use as a model. In spite of financial constraints in recent decades, biological approaches to pest management have been incorporated into crop production systems by the UC Statewide Integrated Pest Management (IPM) Project, which was established by the California State Legislature in 1979. Through this program, the University of California has leveraged modest funding into a highly effective research and extension effort that has earned national acclaim.

There are about 80,000 farmers in California, most of whom rely largely on the advice of about 4,000 licensed pest control advisors in making pest management decisions. Almost every pest control advisor in the state has been reached through the IPM programs. Surveys show that over 90% own at least one of the UC IPM Project's IPM manuals, and about 1,200 of them attend the project's continuing education programs each year.

Since the IPM project began 13 years ago, California farmers have adopted IPM practices in growing numbers; resulting in substantial reductions in pesticide use. For example, a cultural control program for almonds spearheaded by IPM extension staff reduced insecticide sprays for navel orangeworm 43 percent, saving California almond growers an annual $3.6 million in insecticide and application costs. Many more success stories can be cited, including control of caterpillar pests in Sacramento Valley tomatoes, bunch rots in grapes, spider mites in cotton, aphids in Brussels sprouts, and oriental fruit moth and peach twig borer in stone fruits.

Although the initial $1.2 million granted for the base budget of the Statewide IPM Project has not been

One of the many tools the Statewide IPM Program uses is a computerized data bank that contains pest management guidelines for about 700 pests in over 150 situations.

increased since 1983, the program has engendered a large number of projects that are now supported by other government and private sources. It has created a mechanism for identifying and supporting critical short-term research, and developed a delivery system for reaching pest management decision makers in the field. This successful program can be used as a model for future efforts to develop and implement biological approaches to pest management.

In the major findings and recommendations that follow, the Study Group addresses, among other subjects, some funding and institutional issues it regards as basic to the successful development and widespread adoption of biological approaches. In creating new programs, the university's Agricultural Experiment Station and associated Cooperative Extension specialists and advisors should be treated as a state resource. This research and extension structure could, if properly utilized, leverage future allocations of state funds to develop new biological pest management programs as effective and far reaching as the Statewide IPM Program has been.

Increasing concerns about protecting wildlife, environment and water quality have created substantial pressure for the development and adoption of biologically based pest management approaches.

MAJOR FINDINGS AND RECOMMENDATIONS

1.FINDING

Agricultural pest problems have grown rapidly in importance because of the arrival of new pests, the development of new strains and of pesticide-resistant generations, and the evolution of strains of pests and diseases to which formerly resistant crop plants have become susceptible.

This growth in pest control problems is creating problems that could take decades to solve.

In addition, increasing public concerns about environmental quality, the healthfulness of our food supply, and the effects of widespread pesticide use on worker safety and the health of the general public are creating a growing pressure to find safer, more ecologically sound alternatives to synthetic chemical pesticides.

Even as pest problems and public concerns have grown, already inadequate funding for biological approaches to pest control has dwindled further.

Although agricultural pests are always present and evolving, and they continually pose a potential threat, science has temporarily gained the upper hand, largely as a result of the enormous public investment that was made over the last 100 years to develop the present state land grant system with its agricultural experiment stations and extension programs. However, the effectiveness of this system is being slowly eroded across the country.

For more than a century, tax dollars spent on agricultural research have produced some of society's highest net returns, up to 45 percent a year, according to economists. However, for the past three decades, there has been no real growth in funding for agricultural research, and in some areas funding has declined drastically. For example, during the 1980s, the combined effects of inflation and the escalating costs of increasingly complex research resulted in a severe erosion, perhaps 50 to 60 percent in real terms, in the non-salary components of the UC Agricultural Experiment Station budget.

A UC scientist pursues research on tomato resistance to bacterial speck disease. Despite the loss of numerous staff positions, basic and applied research continue.

HERB QUICK

In the area of biological approaches to pest control, funding has never been adequate in California or elsewhere. One reason for this may be that this type of research is by its very nature long-term, and progress is often slow and less dramatic than in other areas of biology.

Recommendation:

California must make an immediate, major, and continuing investment in the development and adoption of biological approaches to pest management as a practical, safe alternative to synthetic pesticides.

2. FINDING

The ability of the University's Division of Agriculture and Natural Resources (DANR) to develop biological alternatives to the use of pesticides has greatly diminished in the past two decades because of the erosion in state support funds. Recent budgetary reductions that targeted organized research significantly

reduced the number of faculty and staff positions, cutting DANR's teaching and research programs. For example, within the Agricultural Experiment Station of DANR, just over 53 academic positions, or 11 percent of the total, were lost between July, 1990 and October, 1991, the bulk of these in pest management and plant science.

At present, DANR does not have the personnel necessary to meet California agriculture's rapidly expanding need for biological approaches to pest control. The numbers of entomologists, plant pathologists, nematologists, crop scientists, animal health scientists, vertebrate ecologists, and those in related disciplines who direct their activities to solving field problems have dwindled markedly in recent years, and those losses have impaired a number of research and teaching areas.

Recommendation:

In recognition of the importance of agriculture to California's economy, and of the potential for serious damage to that economic sector if replacements for chemical pesticides are not found in a timely manner, state allocations should be used to augment the teaching and research faculty and to meet the high-priority needs of California agriculture.

3. FINDING

Cuts in resources and personnel resulting from a greatly reduced state-funded budget and increased reliance on federal grant monies have made it very difficult to maintain a proper balance between basic research and applied research efforts directed toward solving major problems in

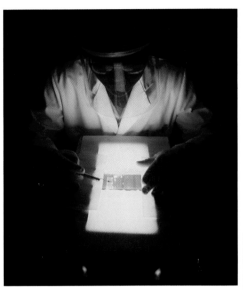

PAUL MIROCHA

A plant pathologist studies DNA fragments fluorescing over ultraviolet light. The fragments form bands according to their sizes, and this information helps scientists isolate and clone genes important in disease induction.

California agriculture. This trend has weakened the ability of the Vice President of DANR to shape the direction of DANR research and to coordinate statewide efforts.

Solving pest problems usually requires an interdisciplinary approach, and often, the personnel and expertise located at one campus or in one county are not sufficient to develop effective programs. Because California is physically large and its agricultural industry is both enormous and complex, the interests of society and of agriculture will be better served to the extent that DANR's Center for Pest Management Research and Extension is successful in coordinating strategies and focusing issues concerned with pest management problems.

Some critics have recently held that UC researchers have not been addressing problems of importance to California. In fact, reductions in state funds have induced many University of California faculty to seek funds from federal grants that focus on basic research, which may or may not have an immediate impact on California's agricultural problems. Because individual researchers have come to rely on extra-mural sources of funds to operate their programs and to support their research assistants, they naturally design research projects in response to their funding sources. Basic research is, of course, a fundamental necessity in any field, and deserves support. However, the reliance of faculty on extramural funding, and the relative increase in the proportion of funds targeted for basic research over those directed toward solving the state's practical problems, has reduced the ability of the Vice President and the Associate Directors of DANR to shape the direction of research.

Recommendation:

DANR should focus attention on present funding patterns and the balance between basic and applied research to design a program that ensures that problems of critical importance to California agriculture are addressed. The Center for Pest Management Research and Extension should be given the resources it needs to coordinate a successful statewide strategy for solving problems in both the near and long term.

In addition, a permanent source of flexible funding should be established to (1) provide research grants designed to stimulate faculty investigations in areas of high priority to the state, and (2) support postdoctorate fellows and laboratory technicians in the performance of such research under the direction of UC faculty. These funding programs would be administered by the Vice President of DANR.

4. FINDING

The communications continuum that extends from the university's basic and applied researchers to its county extension personnel

A graduate student collects aphids from a large color trap as part of research aimed at developing more reliable methods of predicting virus transmission by these insects.

Researchers harvest sugarbeet plots in research designed to evaluate the effects of aphids on the crop. Many of the research programs which growers have adopted widely have involved close collaboration between campus-based researchers and county-based Cooperative Extension advisors.

(who further develop the research) and to the farmers and caretakers of lands (who implement new research in the field) should be strengthened. Existing technology transfer and public education programs are inadequate. California farmers, foresters, and policy makers are not now receiving adequate scientific information about the economic, environmental, and

social costs associated with current and potential pest damage to food and fiber crops.

In Cooperative Extension, DANR has an existing public service and outreach organization that would be well-qualified to meet the state's educational needs in this matter. However, from July, 1990 to October, 1991 alone, Cooperative Extension lost 35, or about 8 percent, of its positions, all specialists or farm advisors.

Because Cooperative Extension personnel are a crucial link between those who do basic research and those who apply it, personnel reductions in Cooperative Extension are hampering field application of research findings, the communication of serious problems in the field to campus-based personnel, and other cooperative programs under which research and extension once flourished.

Furthermore, the development and implementation of sound alternatives to current pest control practices depend not only upon scientific productivity but also upon the effectiveness of educational programs. It is not reasonable to expect California farmers, working alone, to develop innovative solutions to the state's pest control and environmental problems. Nor is it possible for policy makers to generate wise legislation without solid scientific information.

New systems for communicating research developments in pest control are essential. This requires increased emphasis on research and Cooperative Extension activities that provide new information to growers in forms that are readily assimilated and that highlight practical means to adopt new pest management approaches.

Recommendation:

DANR should design and carry out an effective technology transfer program with a strong educational component to encourage growers to adopt effective and profitable biological approaches to pest management as soon as possible.

5. FINDING

Both incentives and disincentives for scientists to develop and for growers to adopt biological approaches to pest management are embedded in the institutional structure of government regulations and laws, private firms who

One IPM program has shown that polyethylene mulch can reduce water use and verticillium wilt in first-year almond and apricot orchards. Some Central Valley farmers are beginning to adopt the practice.

buy and sell farm commodities, and universities.

Growers make decisions about the adoption of pest management strategies in the context of a wide range of institutional forces. For example, federal farm commodity price support programs tend to encourage growers to produce supported commodities (such as cotton and corn) by severely reducing the payments to growers who switch from monocropping to the more diverse crop rotations required for some biological approaches to pest management. In addition, the marketing policies of federal and state agencies and food processors requiring the rejection of fruits and vegetables with superficial scarring are a major financial incentive for growers to apply insecticides prophylactically, according to a predetermined schedule, rather than only when scouting determines there is a risk of loss from pest damage. Special attention should be devoted to the financial risks growers face during the transition period as they experiment with new practices.

Researchers and educators in both the public and private sectors also find that there are pressures inherent in the funding opportunities and incentive structures of their institutions. At the University of California, individual campuses have exerted a powerful influence on the direction of research through faculty appointments and merit and promotion policies. This has, at times, affected the ability of DANR to plan, coordinate, and direct agricultural and environmental research statewide.

Recommendation:

DANR should support research to elucidate the social and institutional barriers to the widespread adoption of biological pest control strategies, as well as research into how to remove them. Potential administrative recommendations include (1) ensuring support for all faculty involved in field-oriented research by granting them 11-month appointments, and (2) encouraging problem-solving research by incorporating it in criteria for academic advancement.

SPECIFIC RESEARCH IMPERATIVES

CHAPTER 1.
Biological Approaches to the Management of Arthropods

Abstract

No species of cultivated plant and no animal escapes injury from arthropods, the animal phylum which includes insects and mites. Most arthropods are harmless to people and agriculture. In fact, many are natural enemies of pests, holding in balance the populations of species that would otherwise become serious pests. Insecticides have been a primary means of battling arthropod pests, but unfortunately, over time, they have engendered many problems, and researchers and farmers are looking to more ecologically sound, biologically based methods for long-term and sustainable control.

There is renewed interest in approaches such as classical biological control, or the importation and release of natural enemies; the release of mass-reared predators and parasites; the development and use of resistant culti-

Intercropping swaths of alfalfa in a cotton field can keep pests out of cotton and provide habitat for beneficial species.

vars; the use of cultural practices, including selective pruning, removal of pests' overwintering sites, and modification of harvesting and planting times; growing insectary plants that nurture natural enemies of pests; use of inter-cropping and trap crops; and the application of more biologically compatible pesticides such as microbials, oils, soaps, and botanicals.

Current basic research promises to open up whole new areas of biologically based arthropod management. Successful use of insect sex attractants (called pheromones) to manage a number of very damaging and previously heavily sprayed insects such as oriental fruit moth and tomato pinworm indicate the great untapped potential of semio-chemicals. New technologies for rearing natural enemies and producing insect pathogens hold much promise for increasing the effectiveness and economic feasibility of these biological agents. Genetic engineering techniques, while they must be used with caution, are likely to yield benefits in improving natural enemy effectiveness, the host range of microbial agents, and pest resistance among crop plants.

Applied research is required to develop and adapt these approaches for many crops and for today's production practices. Effective extension programs are needed to demonstrate the practicality of these types of systems to growers in a clear, convincing manner. For optimal effectiveness, biological approaches must be used within the context of an integrated pest management program. Finally, precise identification of pests and natural enemies to species level is especially important for the effective use of biological methods.

Farmworker releases predatory mites in a strawberry field to control spider mites.

General Goal

Promote basic and applied research to manage arthropod pests through increased use of natural enemies, nematodes and pathogens, behavioral chemicals, cultural methods, and novel field technologies. Such research should improve our taxonomic and systematic knowledge of pests and their natural enemies; generate novel and more effective application methods, formulations and release procedures for biological control agents and semiochemicals, and lead to practical, crop- (or managed-ecosystem-) oriented integrated pest management programs that permit successful use of biological approaches in the context of an overall management program for a multispecies pest complex.

Specific Research Imperatives

1. **Develop more controls of arthropod pests using natural enemies.**

   a. Build, maintain, and staff modern quarantine and insectary facilities to import, rear, and study promising biological control agents.

b. Investigate intrinsic factors that influence the effectiveness of natural enemies.

c. Study host plant relations, ecology, behavior, population dynamics, and crop management practices as they relate to the success of biological approaches.

d. Examine "failed" biological control programs to determine why they were unsuccessful.

e. Evaluate the impact of various pesticides on natural enemies.

f. Improve methods of rearing and maintaining colonies of natural enemies including those of nutrition, quality control, storage, and maintaining genetic resilience and diversity.

g. Improve methods of natural enemy release.

h. Study genetic diversity of natural enemies and acquire knowledge to genetically improve all natural enemy groups.

i. Improve methods of evaluating activities of natural enemies and monitoring the success of biological control programs.

2. **Develop the use of nematodes and pathogens to control arthropod pests.**

a. Search out microbial agents for the control of coleopterous pests, especially those in storage situations.

b. Investigate inheritance of desirable traits of pathogens in order to be able to alter them for specific tasks.

c. Develop *in vitro* mass production methods for economical production of promising pathogens and nematodes.

d. Define the environmental requirements of potentially useful pathogens.

e. Develop strategies to reduce the potential for resistance development with the introduction of transgenic plants.

f. Improve the efficiency of biological control agents through conventional breeding and genetic engineering techniques.

g. Improve formulations, release methods and application technology for microbial agents.

3. **Investigate the use of behavioral chemicals to control arthropod pests.**

a. Chemically identify new semiochemicals for insects that are now, or threaten to be, pests in California.

b. Improve the effectiveness of known blends by identifying previously overlooked but important chemical components.

c. Develop less expensive and more effective methods of synthesizing semiochemicals to increase their economic competitiveness.

d. Improve controlled release technology for emitting effective amounts of semiochemicals in the field over time.

e. Develop pheromone mimics to improve the stability and commercial availability of semiochemicals that have complex or unstable structures.

f. Determine the critical aspects of the life cycles of pest species to ascertain the stages and behaviors that are most vulnerable to behavioral manipulation through semiochemicals.

g. Work with regulatory agencies to improve and streamline registration procedures.

h. Educate the public about the safety of semiochemicals.

4. **Carry out more field-oriented research in practical situations.**

a. Increase studies on the field ecology of pests, their hosts, and natural enemies with special consideration for inter- and intraspecies interactions under current and future production practices.

b. Encourage research to demonstrate, evaluate, and improve utilization of alternative techniques in practical situations.

c. Improve and expand field stations with equipment and staff to encourage field-oriented research and teaching.

A UC researcher uses a vacuum collector in a strawberry field to evaluate the presence of a parasite of lygus bug pest.

d. Make allowances for the long-term commitments frequently necessary for completing research and achieving success in the complex field environment.

e. Strengthen Cooperative Extension to improve the link between research and consumer.

CHAPTER 2.
BIOLOGICAL APPROACHES TO WEED MANAGEMENT

Abstract

Weeds can cause severe economic, environmental and human health impacts over a wide range of agricultural and nonagricultural environments. Despite growers' efforts, weeds are estimated to reduce California crop yields 4 to 20 percent, depending on the crop, causing an estimated billion dollars in yield reductions annually. Currently, herbicides are the primary tool used to combat weeds, making up over 60 percent of the pesticides sold in California. Machine and hand cultivation are also used extensively in agriculture to eliminate weeds.

Biological methods of weed control are based on the use of other organisms to stress weeds through (1) plant feeding by biological control organisms (herbivory) and (2) plant competition. In "classical" (or "inoculative") biological control, natural enemies of weeds are imported from the weed's area of origin. (Most severe weeds are of foreign origin, having left their natural enemies behind.) Research involves foreign exploration for natural enemies, assessing their host-specificity and safety for release, importing and establishing natural enemies as biological control agents, and post-release evaluation studies. Arthropods (chiefly insects) and pathogens (chiefly fungi) include many host-specific natural enemies sufficiently safe to import as biological control agents. In the "augmentative" or "inundative" approach to biological control, weed pathogens can also be mass-produced in culture and large amounts of inoculum (usually spores) applied to a weed target as a "bioherbicide."

The plant competition approach is based on the use of other plants to impose stress on neighboring weeds by depleting limited resources, usually light, or through the competitive plants' release of chemicals that impair weed growth (allelopathy). In cropland environments, this usually involves the use of cover crops or smother crops. Crops may also be genetically improved for enhanced competitiveness with weeds.

General Goal

Develop the scientific capability to utilize fully the potential of arthropods, pathogens, nematodes, allelopathy, and cultural practices to control weeds.

Specific Research Imperatives

1. **Develop and improve the use of arthropods for biological control of weeds.**

   a. Discover new arthropod biological control agents for introduced weeds through foreign exploration in the areas of origin of weeds.

   b. Assess the host specificity of biological control arthropods through field and laboratory studies, and elucidate the determinants of host specificity.

   c. Assess through post-release field evaluation studies the status of biological control arthropod populations and their impact on target weeds.

   d. Develop methods to integrate the use of arthropod agents in cropland environments with reduced levels of pesticides and soil tillage.

   e. Investigate the genetics of cases of apparent adaptive microevolution following agent introduction and establishment.

   f. Determine the impact of genetic variability

Cinnabar moth (*Tyria jacobaeae*) larvae (left) feed on the flowerheads of tansy ragwort. The larvae can completely strip the weed of flowers and leaves (right), strongly limiting seed production. The moth was imported from France to help control the weed.

of introduced arthropods on their effectiveness as biological control agents.

g. Develop methods of genetic enhancement of arthropod biological control agents.

2. Develop and improve the use of pathogens for biological control of weeds.

a. Discover new biological control pathogens for introduced weeds through foreign exploration for pathogen agents in the areas of origin of weeds.

b. Assess the host specificity of biological control pathogens through laboratory studies in domestic quarantine facilities.

c. Devise methods to increase the susceptibility of weeds to infection and to improve weed infection by applied pathogen inoculum through the development of new adjuvants to facilitate inoculum survival and infectivity.

d. Develop mechanisms to foster public-private sector cooperation and commercial development of pathogens ("bioherbicides") used in inundative releases.

e. Develop methods of genetically enhancing pathogens used as biological control agents to increase spore and/or toxin production, and to increase host specificity and degree of environmental persistence (increase persistence in highly host-specific pathogens

The larvae of the ragwort fleabeetle (*Longitarsus jacobaeae*), shown here in its adult form, kill weeds by feeding on the roots. The beetle was imported from Italy.

and decrease persistence in less specific pathogens).

3. Develop and improve methods of using plant competition to suppress weeds, and increase crop competitiveness with weeds.

a. Elucidate the nature and dynamics of interspecific plant competition, including both resource competition and allelopathy.

b. Develop practical methods of increasing the use of cover crops and smother crops to suppress weeds in cropland environments.

c. Develop practical methods of increasing plant competition on weeds of pastures and rangelands through improved grazing man-

agement and/or the seeding of environmentally benign plants competitive with weeds.

d. Identify potentially valuable germ plasm for increased competitiveness among the wild relatives of crops.

e. Develop screening programs for identifying more competitive crop germ plasm in complex, weed-infested cropland environments.

f. Genetically enhance crop competitiveness with weeds while ensuring that the genetically enhanced crop itself does not become a weed or transfer traits enhancing weediness to interfertile wild relatives.

4. **Improve our knowledge of the taxonomy and systematics of weeds and their natural enemies.**

a. Develop efficient methods of determining the source areas of introduced weed biotypes in order to search more efficiently for effective natural enemies.

b. Elucidate the taxonomy of important groups of arthropod natural enemies.

5. **Elucidate the ecological underpinnings of biological control systems.**

a. Elucidate the dynamics of weed populations, natural enemy populations, and their interactions.

b. Elucidate the factors affecting the susceptibility of environments to weed invasions.

c. Elucidate the factors fostering the support of healthy populations of established natural enemies in agricultural and nonagricultural environments.

CHAPTER 3.
Biological Approaches to the Management of Plant-Parasitic Nematodes

Abstract

Nematodes can be both crop pests and the natural enemies of pests. Nematode populations in soil and water may require management for their impacts, both negative and positive, on the environment, on plants, on animals, and on humans. There are intriguing possibilities for nematode management strategies that do not rely on pesticides. They emerge from studies in nematode ecology, host-parasite interactions, and from advances in the understanding of the molecular genetics and molecular biology of nematodes.

Considerable developmental and adaptive research will be necessary to facilitate the marriage of basic biology and ecological theory with agricultural, environmental, and economic reality. There are several important avenues of development. Natural products chemistry and chemical ecology offer opportunities in attraction, repulsion, sensory confusion, and life-cycle disruption of nematodes. Population regulation

Before biological control a Mendocino County ranch was infested by tansy ragwort (left). The same ranch is shown (right) after the introduction of the cinnabar moth and the ragwort fleabeetle reduced the weed to less than 1% of its former population.

through exploitation, competition, and antibiosis by antagonists is another area to be further explored. Host genotypes that exhibit resistance or tolerance to parasitic species provide basic genetic material for traditional and transgenic approaches to protecting plant and animal species.

At a population and community level, there is great potential for innovation in ecosystem design through cultural management, and through temporal and spatial arrangement of host and nonhost species. Throughout these approaches lie challenges, not the least of which is the need for rapid and reliable target-specific diagnostic tools. This chapter documents the state of the existing knowledge and offers analysis of some promising directions. It outlines the research and resource needs to allow the new approaches to be implemented. Some are ready for field testing; others require more extensive development. However, in almost all cases, the direction and path are clear.

General Goal

Develop effective approaches for the biological control of nematodes.

Specific Research Imperatives

1. **Develop and implement management tactics involving chemical ecology.**

   a. Evaluate effects of plant root exudates in creating a rhizosphere unfavorable for nematodes.

   b. Design cropping systems to incorporate the effects of allelopathic plants.

   c. Identify materials that attract, repel or disrupt nematode behavior through sensory confusion.

   d. Determine the function of sensilla in host-finding and mating behavior, and their response to repellent and attractant stimuli.

2. **Study the biological mechanisms of nematode antagonism to determine where and**

UC farm advisor examines a wheat root for presence of Columbia root knot nematode.

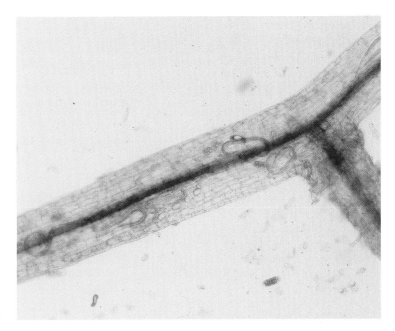

Root lesion nematodes are pests of numerous California crops.

when biological control is likely to be effective.

a. Develop quantitative and qualitative assays for nematode antagonists and soil suppressiveness.

b. Develop DNA probes and other molecular markers for identification of nematode antagonists.

c. Develop tactics for introduction, inundation and augmentation of biological control agents of soil-inhabiting nematodes.

3. **Develop plant cultivars that are resistant to nematodes by using conventional breeding and genetic engineering methods.**

    a. Identify and develop natural and novel gene sources for resistance to plant-parasitic nematodes.

    b. Implement classical and novel gene transfer techniques for development of resistant cultivars.

    c. Elucidate the underlying mechanisms of nematode and plant interaction, and identify genes and gene products that will reduce the vulnerability of crop plants to nematode damage.

    d. Determine the potential for immunization of plants against nematode parasitism.

    e. Study the transfer and expression of foreign genes in plants, including those from microorganisms that control production of microtoxins.

    f. Determine the range of genetic diversity within species of plant-parasitic nematodes to allow management of resistance genes.

    g. Develop a database of cultivars of crops resistant and tolerant to the nematode species and races prevalent in California.

4. **Devise methods of controlling nematodes through physical and cultural means.**

    a. Develop cultural practices to control specific nematodes in different crop and biogeographic systems.

    b. Study the life history and survival strategies of plant-parasitic nematodes to determine their vulnerable stages.

    c. Incorporate planting and harvest date adjustment, crop rotation, use of cover crops, and physical methods into cropping systems.

5. **Study the ecology of plant-parasitic, bacterial-feeding, fungal-feeding, and predaceous nematodes.**

    a. Determine the role of nematodes in cycling of mineral nutrients and as components and indicators of soil health.

    b. Determine the contribution of nematodes in the microbial decomposition of pollutants in soil and water.

6. **Improve our understanding of nematode biodiversity.**

    a. Develop diagnostic techniques for rapid and accurate identification of nematode species and biotypes for management decisions.

    b. Develop nematode taxonomy and document intraspecies genetic diversity.

7. **Increase UC Cooperative Extension activities in applied nematology.**

    a. Add personnel in Extension training and advisory services.

    b. Increase activity of experiment station personnel in applied research and implementation.

    c. Develop expert systems and other guides for management decisions in specific crop and biogeographic situations.

CHAPTER 4.
Biological Approaches to the Control of Plant Diseases

Abstract

Plant diseases are a major cause of crop damage in California, destroying an estimated 10 percent and more of each of the crops grown in the state. This results in significant economic losses. Some diseases pose threats to human health. California growers rely heavily on chemical pesticides to control many types of plant diseases. This dependence will continue until effective alternative methods of disease control are developed or pesticide use is banned.

Existing nonchemical methods of controlling plant disease include cultural

MILTON SCHROTH

Growers can now control crown gall disease in stone and pome fruits by using a harmless strain of the same disease to colonize the plant roots. Roots on right have been protected with this strain.

control, exclusion, and plant resistance and immunization. These methods are also included in the strategies and tactics of integrated pest management. Though biological control methods have proven effective against a number of insect pests, mites, and weeds, they are less well-developed in controlling plant diseases. Research is essential to strengthen this area of plant pathology. It is also necessary to extend our knowledge in proven areas of disease control that are environmentally safe but on which little applied research has taken place during the past three decades. New research on improved cultural practices and plant resistance control methods could offer meaningful alternatives to chemical pesticides as a primary means of disease control. Genetic engineering and other biotechnological methods now offer many novel approaches to controlling pests.

General Goal

Promote basic and applied research aimed at developing new and safer approaches for managing plant diseases of agricultural and forest plants with emphasis on nonchemical methods.

Specific Research Imperatives

1. **Develop new and improved methods to control pathogens biologically.**

   a. Identify antagonists of major plant pathogens and determine factors that increase their effectiveness as biocontrol agents.

   b. Develop methods to identify and monitor the activities of key pests and their antagonists.

   c. Determine mechanisms by which antagonists suppress populations of pathogens.

   d. Determine ecological and management factors that affect the survival dynamics of biocontrol agents.

   e. Develop technologies for storing and delivering biocontrol agents to target sites.

UC scientists are developing methods of rapid disease detection and monitoring that will enable growers to detect plant diseases such as phytophthora root rot before symptoms become obvious.

2. **Improve physical methods of controlling pathogens such as soil solarization, flooding, and soil tillage depositing propagules at disease-inhibiting depths.**

3. **Improve health of agricultural and forest plants by crop rotation and management practices.**

   a. Determine environmental factors that discourage the survival and activities of key pathogens in California.

Applying IPM-generated research, grape growers use such equipment as this mechanical pruner to reduce bunchrot in grapes through canopy management.

b. Determine the effects of soil moisture and structure, cover crops, tillage, and other management practices on populations of pathogens and beneficial organisms.

c. Determine the environmental factors that increase the susceptibility of plants to attack by pathogens.

d. Discover vulnerable stages in the life cycles of plant pathogens for devising control strategies such as host-free periods.

4. **Improve the safety of pesticide use and develop better methods of application. Devise methods and develop environmentally driven models to apply pesticides more effectively and lessen their spread into the environment.**

5. **Develop and improve integrated pest management (IPM) strategies for managing plant diseases.**

a. Develop better monitoring of pathogen activities and their role in decreasing yields.

b. Integrate research with other disciplines leading to the development of complete management systems.

6. **Improve crop plant disease resistance by classical breeding and genetic engineering methods.**

a. Determine the genes and products that enable pathogens to attack hosts and the concomitant genes for resistance.

b. Search for resistant germ plasm.

c. Develop novel methods of controlling diseases such as genetic altering to support a microflora antagonistic to key pathogens.

d. Develop improved strains of biological control organisms.

e. Develop improved methods to immunize the host by cross protection, induced resistance, and molecular parasitism.

f. Determine the basis for plant immunization.

CHAPTER 5.
Biological Approaches for Controlling Food Animal Pests

Abstract

If they are not prevented or controlled, animal diseases cause enormous economic loss through death, culling, reduced productivity, and treatment costs. Since food animal commodity sales (largely milk, meat, and poultry) represent about 29 percent of California agricultural sales, these losses are significant.

Some animal disease pests can cause disease in humans and a variety of microbial and parasitic pests can contaminate foods of animal origin. Misuse of chemicals intended to control animal pests and animal ingestion of feedstuffs containing natural or synthetic chemicals may leave harmful residues in foods of animal origin.

Many of the chemicals that have been used to control food animal pests are being withdrawn. Some are becoming ineffective through the genetic resistance of parasites and other pathogens. Other chemicals have been lost through regulatory action linked to the risks of residues in food animal products. Since food animal products may be less safe following a chemical ban because of the increased risk to both animal and human health from animal pests that could proliferate following such a prohibition, it is imperative to develop alternative strategies to treat and control food animal pests. Alternatives to withdrawing effective chemical controls include developing precise drug monitoring programs, controlled medications, and effective, efficient screening for drugs and chemicals prior to slaughter or sale for human consumption.

The biological alternatives for controlling animal pests are more complicated than those for plant diseases, but they are essential if chemical controls are withdrawn or become obsolete due to pest resistance. Expanding existing biological approaches to animal pest control and developing others will require an integrated effort. This should involve working to eradicate and exclude select-

California's vast food animal populations are at risk from a large variety of pests, including viruses, bacteria, and internal or external parasites. Milk and milk products are the state's number one agricultural commodity, with production valued at $2.15 billion in 1989.

ed diseases, altering the food animal environment to make it less suitable for disease agents and their vectors, adopting high standards of animal care and strategic vaccination programs, developing natural biological response modifiers, and carefully selecting or, where feasible, engineering animals that are resistant to infection. These approaches may be very expensive. So far, only one disease, screwworm, or miasis, has been prevented by biological control, and a continuous international effort is required to avoid reintroducing screwworm into the United States. Major basic and applied research commitments are needed to improve the effectiveness of these efforts.

General Goal

Discover effective and economically feasible ways to minimize chemical use in food animal populations and chemical residues in food animal products.

Specific Research Imperatives

1. Develop a comprehensive understanding of the diseases affecting our food animal species in order to devise techniques to exclude, eradicate, or control disease in food animal populations.

2. Develop and adopt environmental design features and animal husbandry methods that minimize the occurrence, intensity, and spread of disease in our food animals.

3. Identify the critical sites where microbial and chemical agents enter the human food chain in animal products, and establish quality control programs that intervene before contaminants enter human foods.

4. Model and fully examine the effects of various drug control policies on animal and human health before they are instituted.

5. Thoroughly characterize human exposure to animal drugs and any threat they may pose to human health.

6. Develop new vaccines which enhance food animal immunity to disease.

7. Develop naturally occurring substances and no-residue chemicals which enhance the host's resistance to disease.

8. Improve procedures for diagnosing diseases and for administering medication. Develop better methods to identify ani-

mals that have been treated with medicine in order to ensure that appropriate therapy is given and that full drug withdrawal times are followed before slaughter.

9. Identify and select animals with disease resistance traits using conventional breeding and genetic engineering techniques.

10. Devise systems to ensure that the producer, veterinarian, chemical manufacturer, food retailer, risk assessor and consumer are informed and educated as new disease control facts are recognized. Encourage the adoption of techniques acceptable to society that optimize both production and safety of foods of animal origin.

CHAPTER 6.
Ecological Approaches for Controlling Vertebrate Pests

Abstract

Native and exotic vertebrate pests can cause substantial economic losses in farm, ranch and forest agricultural systems. For example, blackbirds cause an estimated $34.8 million damage annually to corn nationwide.

Biological approaches to the control of vertebrate pests emphasize removing the impact of a pest species rather than eradicating the pest population. These methods must be based on a knowledge of the biology and ecology of pests and their antagonists. Mechanisms that regulate populations must be determined. Control can then be achieved by regulating these mechanisms over time. The spatial scale at which control must be applied to be effective must be determined. Scientific studies of population dynamics and species ecology can establish general approaches which can be rapidly applied to given local conditions. While general knowledge of population principles is well-established, scientists sorely lack the necessary detailed knowledge of the functioning of ecosystems and how natural or human-induced perturbations affect those systems, including component commodities and pests.

General Goal

Quantify the influence of abiotic and biotic factors on population numbers of target species and their predators and competitors.

Specific Research Imperatives

1. Determine factors influencing long-term population density of vertebrates.

   a. Improve our understanding of the interactions between an organism and its environment, concentrating on physiological responses to natural and human-induced stresses.

California's production of chickens (left), turkeys (right) and eggs was valued at $1.01 billion in 1989.